THE ZAMBIT TRIANGLE DREAM

KlevaKids

Published by KlevaKids.com Inc

New York, New York

For more titles,visit our website at www.klevakids.com

KlevaKids

KlevaKids

Presents

THE ZAMBIT TRIANGLE DREAM

This story introduces the basic types of triangles in an enjoyable format.

The above statements are purely informational. The purpose of the book is not to teach, but to expose concepts. Read the book to a child and just allow the story to be enjoyed.

There was that unruly

pile of stuff.

It made me feel

like I wanted to huff.

The math symbol contraption came

with quite a bit of zing.

Surely it could give order

to this unruly thing.

The contraption moved in

with such grace and ease.

Making order of this stuff

was such a breeze.

Several line segments

shot out of the top.

They came out neatly

with a swoosh and a pop.

When the contraption left

it was appreciated truly.

All that stuff

was no longer unruly.

It was moving slowly

from out in the distance.

A geometry ship came

to provide assistance.

It had three math Zambits

and I could hardly wait.

I knew that what they'd show me

was really going to be great.

Three Zambits connected line segments,

all of the same size.

It made an equilateral triangle.

Now that was quite a surprise.

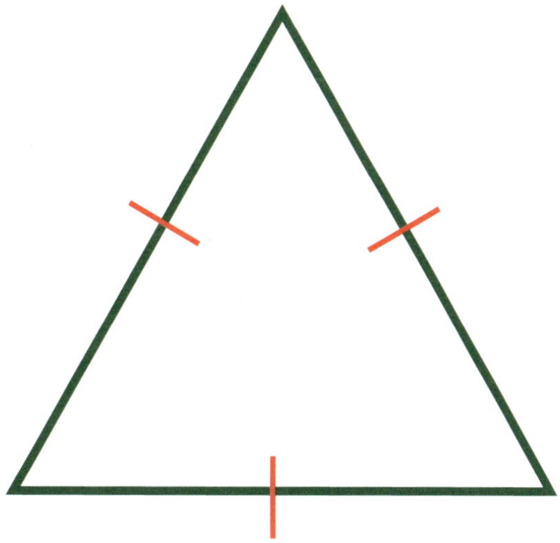

Equilateral Triangle

All sides of equal length

They got three more line segments.

Only two of equal length were selected.

It made an isosceles triangle

when they were all connected.

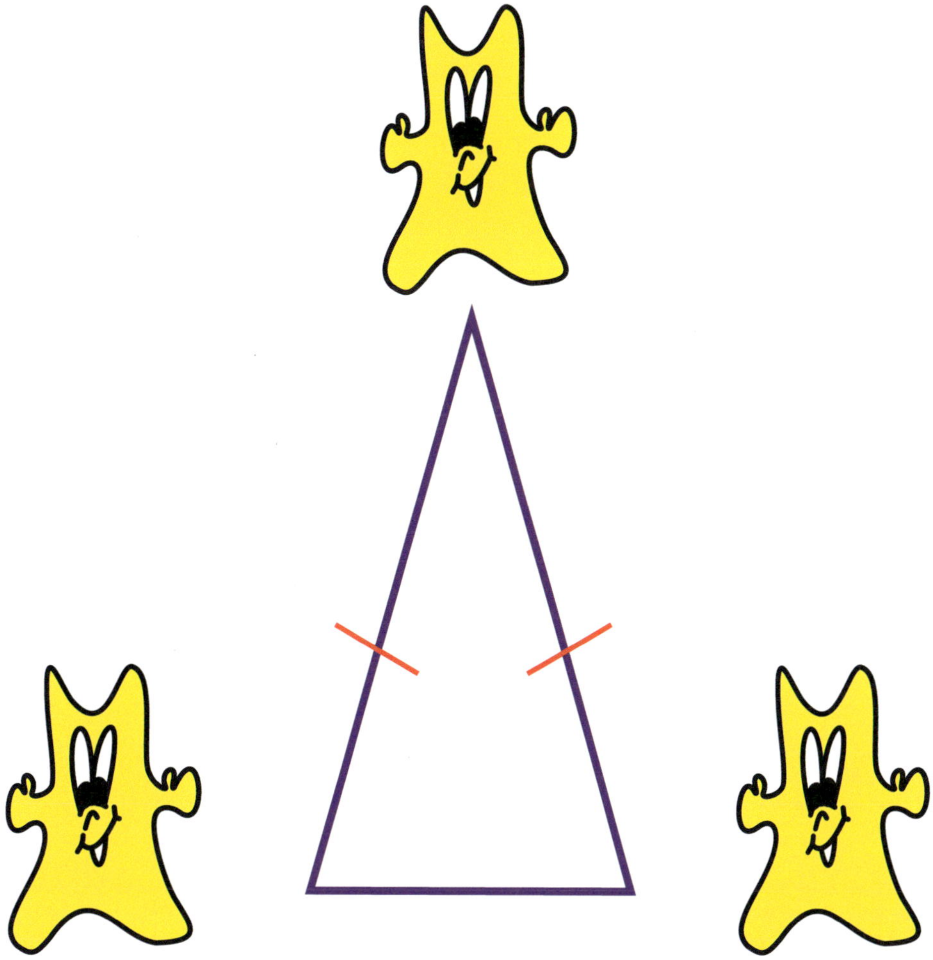

Isosceles Triangle

Two sides of equal length

Three other line segments were connected with no lengths the same. They made a scalene triangle and that was their aim.

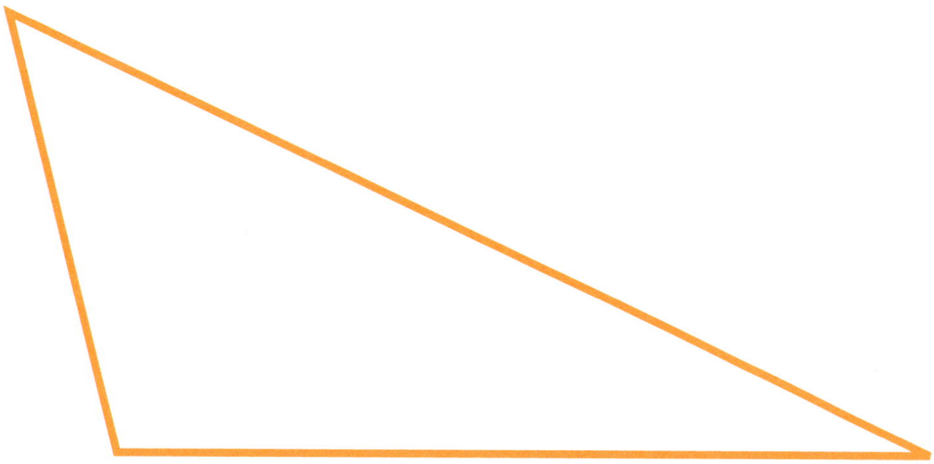

Scalene Triangle

No sides of equal length

Line segments were connected

with one angle ninety degrees.

Those Zambits made a right triangle

and knew how to please.

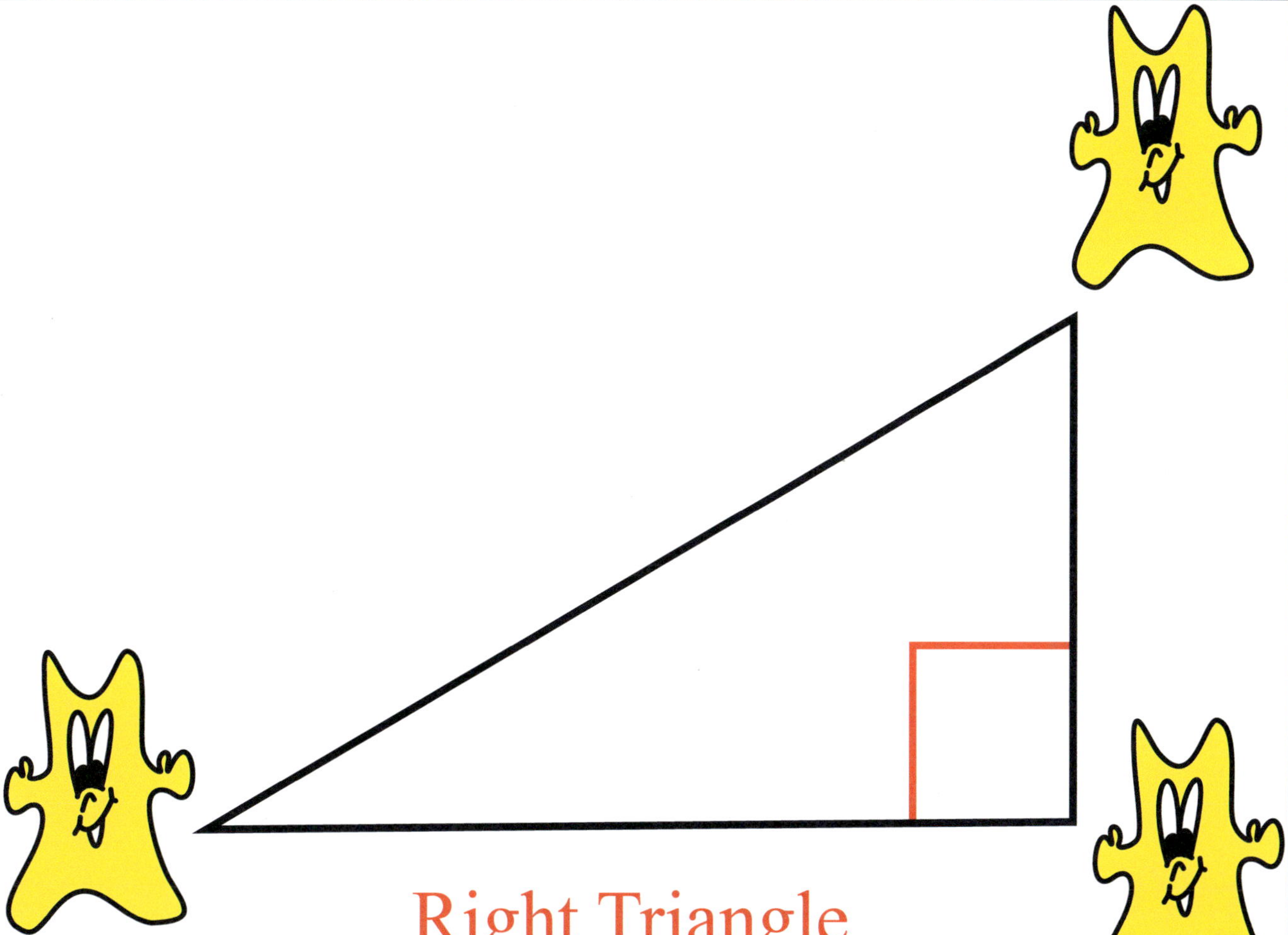

Right Triangle

One angle equal to 90 degrees

Surely no other line segments

could have much use,

but one angle greater than ninety degrees

made the triangle obtuse.

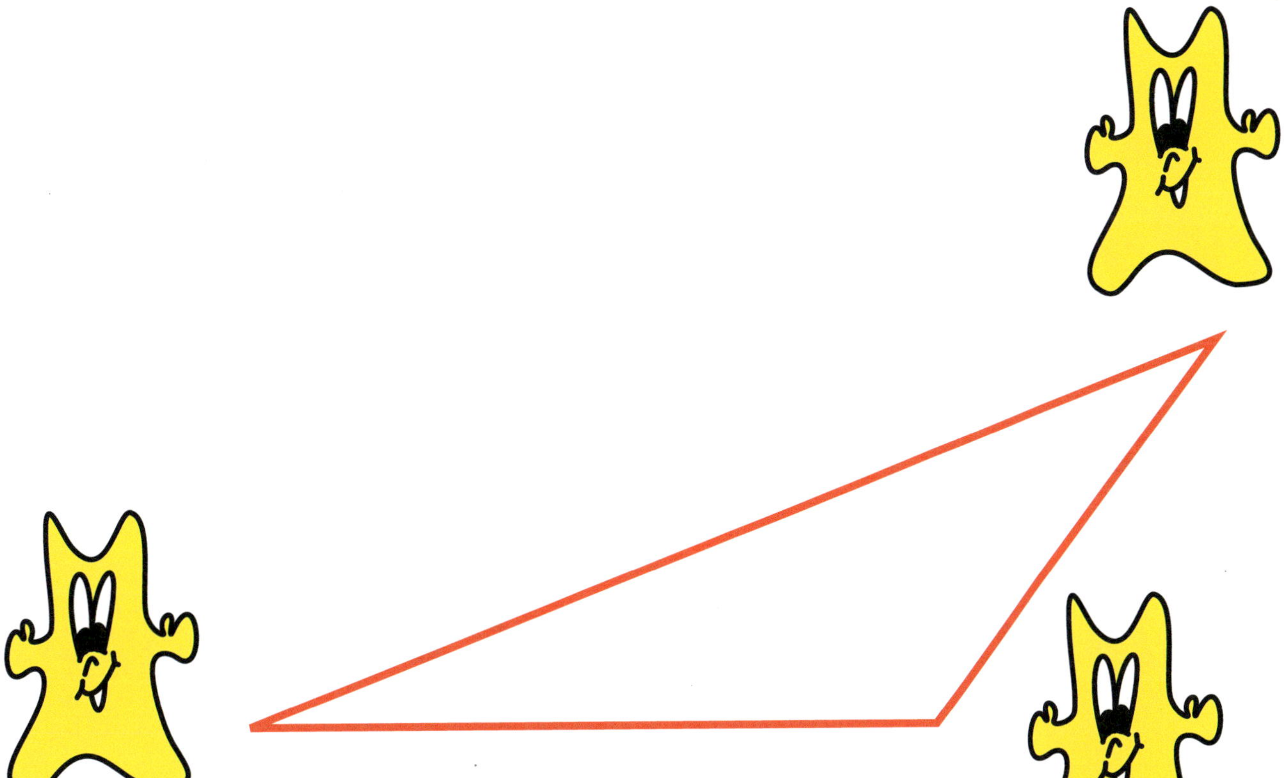

Obtuse Triangle

One angle greater than 90 degrees

The Zambits had a finale,

it made me want to hoot.

Every angle was less than ninety degrees

and they called the triangle acute.

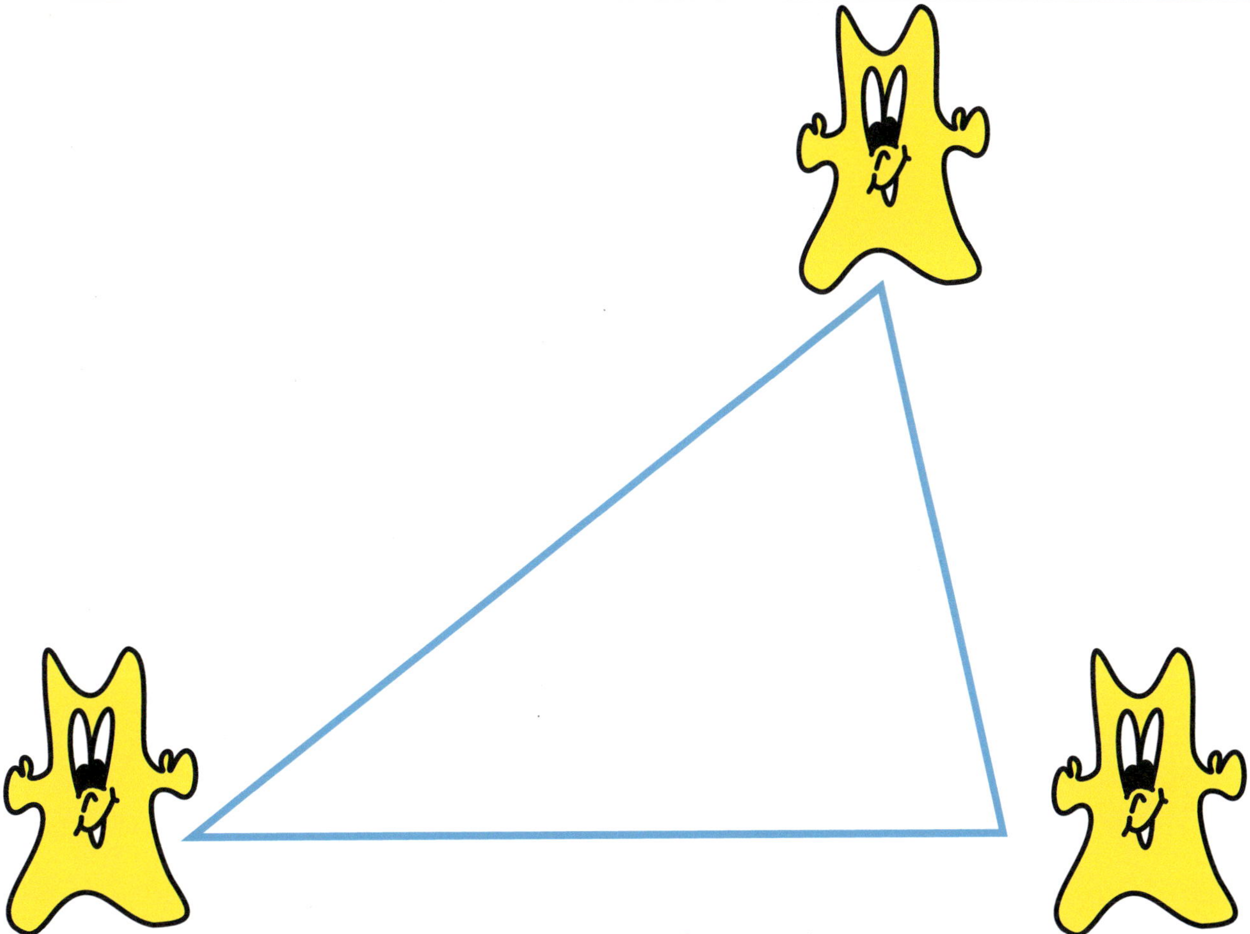

Acute Triangle

All angles less than 90 degrees

The Zambits had to leave.

I knew that from the start.

I'm just glad those Zambits

are really very smart.

I love to dream of Zambits.

That's exactly what I said.

I like to keep this triangle stuff

right inside my head.

THE END

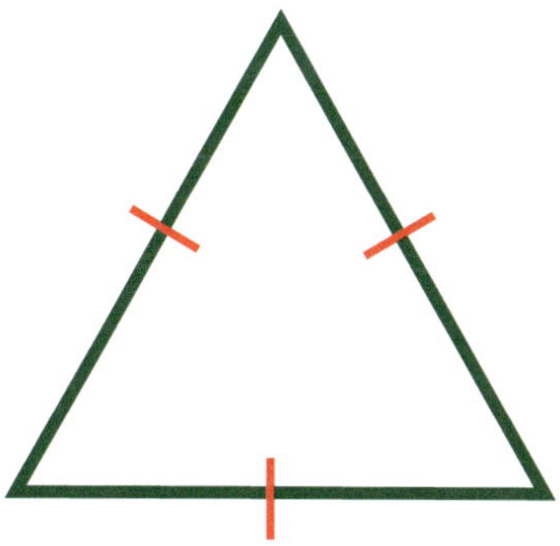

Equilateral Triangle
All sides of equal length

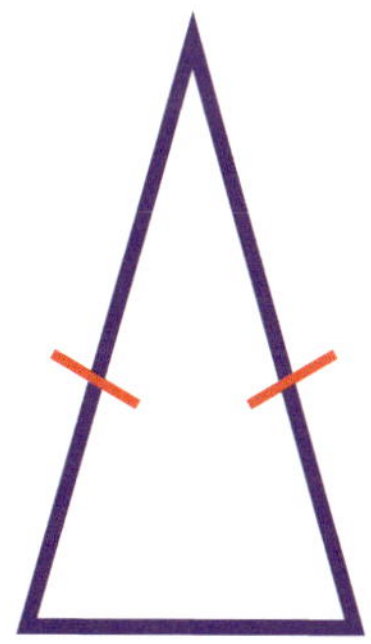

Isosceles Triangle
Two sides of equal length

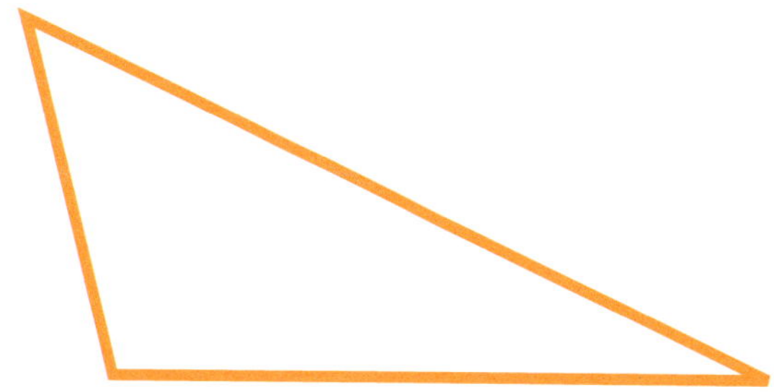

Scalene Triangle
No sides of equal length

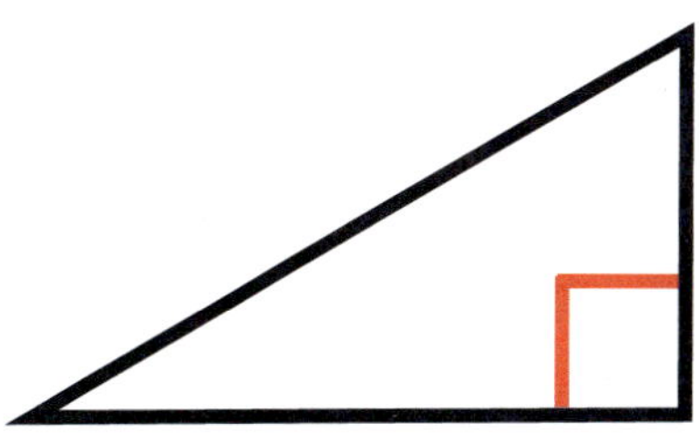

Right Triangle
One angle equal to 90 degrees

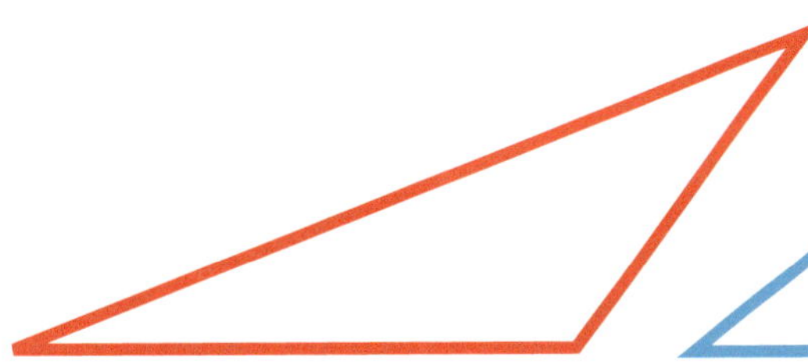

Obtuse Triangle
One angle greater than 90 degrees

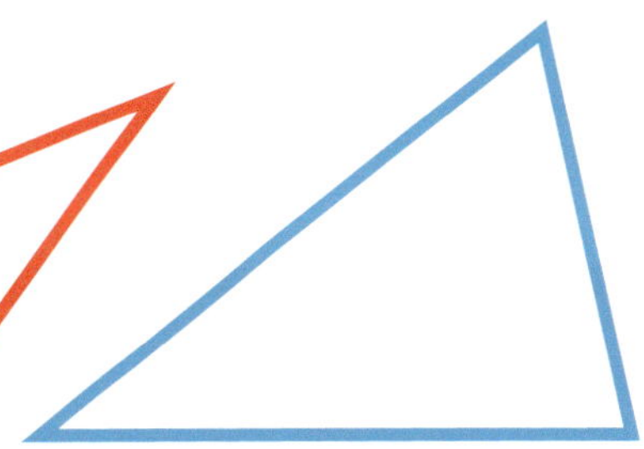

Acute Triangle
All angles less than 90 degrees

Made in the USA
Monee, IL
12 September 2026

62974509R00021